YOUR KNOWLEDGE HAS VALUE

Bibliographic information published by the German National Library:

The German National Library lists this publication in the National Bibliography; detailed bibliographic data are available on the Internet at http://dnb.dnb.de .

Imprint:

Copyright © 2020 GRIN Verlag
Print and binding: Books on Demand GmbH, Norderstedt Germany
ISBN: 9783346213082

This book at GRIN:

https://www.grin.com/document/914160

Anonym

Aus der Reihe: e-fellows.net stipendiaten-wissen

e-fellows.net (Hrsg.)

Band 3504

The Architecture of Convnets and Data Processing. Advantages of Convolutional Neural Networks

GRIN Verlag

GRIN - Your knowledge has value

Since its foundation in 1998, GRIN has specialized in publishing academic texts by students, college teachers and other academics as e-book and printed book. The website www.grin.com is an ideal platform for presenting term papers, final papers, scientific essays, dissertations and specialist books.

Visit us on the internet:

http://www.grin.com/

http://www.facebook.com/grincom

http://www.twitter.com/grin_com

Mathematical Introduction to

Machine Learning

Essay

Convolutional Neural

Networks

Course of studies: **Master in Business Mathematics**

Content

1   Introduction

In the past two decades in particular, artificial neural networks have led to new approaches and processes in machine learning in many areas. They have replaced many existing processes. In some areas, they even exceed human performance. Impressive progress has been made in the area of image recognition and classification. Above all, this includes the introduction of convolutional neural networks (ConvNets). They belong to the class of neural networks. The first ConvNet was developed by LeCun et al. in 1989. [1] ConvNets were especially developed to enhance image processing. Therefore, they provide a unique architecture. Due to their structure and functionality, ConvNets are particularly well suited within this field of application compared to other methods.

1.1   Motivation

Ordinary neural networks basically consist of an input layer, a series of hidden layers and an output layer. Every single hidden layer consists of a number of neurons. Each neuron is connected to any other neuron of the preceding and following layer. The neurons operate independently and do not share any weights with other neurons of the same layer. The output layer represents the results, for example the class scores in case of classification tasks. Although impressive results in many applications using classic neural networks are achievable, they are not suitable for all scenarios due to their rather simple structure.

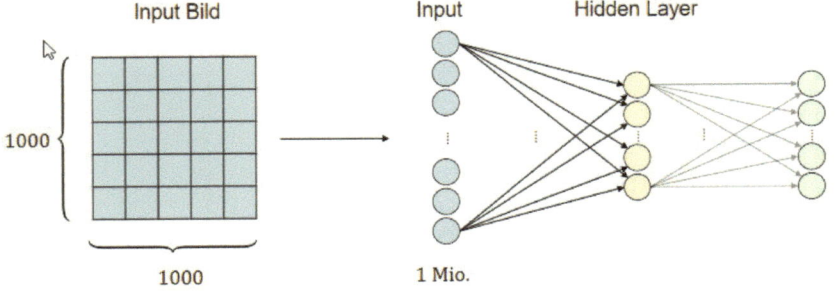

Figure 1: Image processing in ordinary neural networks[2]

[1] Le Cun, Y., Boser, B., Denker, J. S., Henderson, D., Howard, R. E., Hubbard,W., & Jackel, L. J. (1989). Backpropagation applied to handwritten zip code recognition. Neural Computation, 1, 541–551

[2] Nikolić, Zoran: (2019). Convolutional Neural Networks, URL: http://www.mi.uni-koeln.de/wp-zni-kolic/wp-content/uploads/2019/06/11-Odenthal.pdf, June 22, 2020

For example, consider a record of 1000 x 1000-pixel images that have to be classified using an ordinary neural network. Each pixel of an image corresponds to a neuron of the input layer. Therefore, the first layer already consists of one million neurons. Hence, for any additional neuron in the first hidden layer one million edge weights would have to be recalculated in each training step. In case of colored images, that issue of computational effort becomes even more acute, because each pixel is mapped into a three-dimensional color space. With respective magnitude, a normal computer quickly reaches its limits. ConvNets try to compensate this disadvantage. ConvNets are similar to ordinary neural networks. They have an input layer, a certain number of hidden layers and an output layer. Furthermore, layers in ConvNets are also equipped with neurons, weights and biases. The main difference to neural networks is related to the architecture and the way the first few hidden layers work. In contrast to neural networks, most layers in ConvNets are only locally connected to preceding layers. This significantly reduces the training effort. The following section explains the individual components of a ConvNet and its way of processing data.

2   The Architecture of ConvNets and Data Processing

ConvNets essentially consist of filter layers (called *convolutional layer*) and aggregation layers (called *pooling layer*), which are repeated alternately. They are followed by one or more fully connected layer.

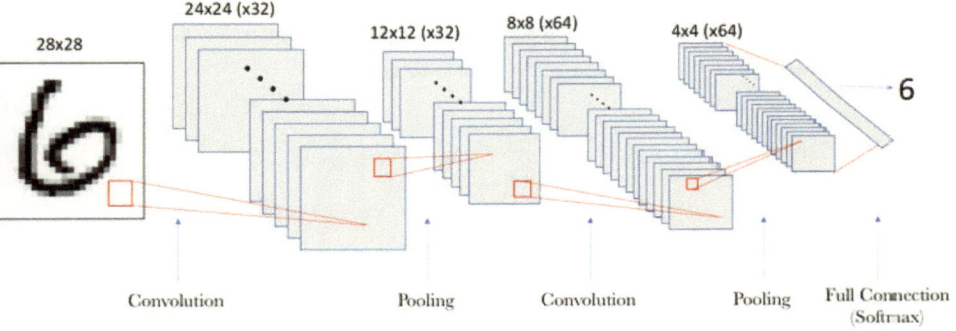

Figure 2: An example ConvNet architecture[3]

In the following, the case of grayscale images is dealt with first. In order to be able to process colored images with ConvNets, only minimal changes have to be made. However, the general structure and its individual processing steps remain the same.

2.1   The Convolutional Layer

At first, the matrix input is analyzed by a predefined number of *filters* (also called kernels) of a fixed size. While processing, they move like a window with constant step size (called *stride*) over the pixel matrix of the input. The filters move from left to right over the input matrix and jump to the next lower line after each run. *Padding* determines how the filter should behave when hitting the edges of the matrix.

[3] Torres.AI, Jordi: (2018). Convolutional Neural Networks for Beginners, URL: https://towardsdatasci-ence.com/convolutional-neural-networks-for-beginners-practical-guide-with-python-and-keras-dc688ea90dca, June 22, 2020

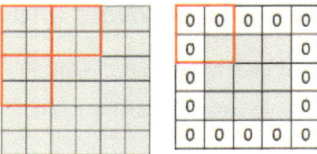

Figure 3: Stride and Padding[4]

If padding is used, a margin of zeros is added around the original matrix. While processing, this allows the original size of the input to be retained.

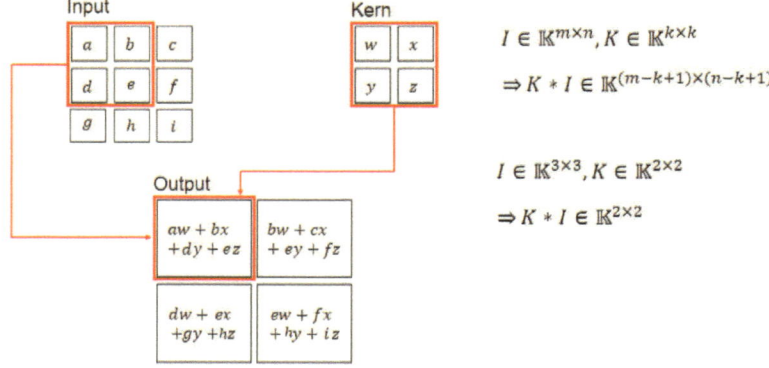

Figure 4: The convolution operation[5]

The filter has a fixed weight for every point in its viewing window. The weights do not change when running through the initial input matrix. As result, the *feature maps* are calculated by *convolution*. The operation of convolution consists in performing point products between the filter weights and the local image section values and adding them up afterwards. Given an m x n input image I and a filter K of dimensions k_1 x k_2, the discrete 2D-convolution at point (i, j) is defined by:

$$(I * K)_{i,j} = \sum_{m=0}^{k_1-1} \sum_{n=0}^{k_2-1} I_{i-m,j-n} K_{m,n}$$

[4] Based on: Nikolić, Zoran (2019)
[5] Based on: Nikolić, Zoran (2019)

4

The dimensionality of the output matrix depends on the image size, filter size, the padding (p) and the stride (s).

$$Output \in K^{\left(\frac{m+2p-k_1}{s}+1\right)x\left(\frac{n+2p-k_2}{s}+1\right)}{}_{67}$$

For example, a stride of 2 in combination with a filter size of 2 x 2, results in a quarter of the dimensionality of the input matrix. The output of the convolutional layer serves as input for the subsequent pooling layer.

2.1.1 Hyperparameters and filter weights

The hyperparameters of a ConvNet include the number of filters, the size of the filter windows, the stride and padding. The number of filter blocks can be interpreted as the number of features to be treated. It's a power of two anywhere between 32 and 1024 and is usually set to 32 or 64. The filters dimensionalities and their weights are not fixed and by convention often set to a small odd number depending on the application.[8] One general approach is to use larger filters on high dimensional data and smaller filters when considering opposite cases. Another approach is to start using small size filters and to gradually increase them in subsequent layers. The filter weights are chosen randomly at the beginning and become further optimized during the training. Non-positive values are also approved for assignment. Stochastic gradient descent in combination with back propagation is commonly used as training method. By minimizing the loss function, the optimal (or at least locally optimal) set of weights can be found.

As mentioned in a previous section, padding is used to prevent shrinking dimensions of output matrices by adding additional zeros around the input images before sliding the filter window through it. It can be considered as a trade-off between information loss and reduced dimensionality. [9]Another mentioned hyperparameter of a convolutional layer is the stride, which indicates the number of pixels that the window moves in each step. [10]Like Padding, there is a trade-off between information loss and dimensional

---

[6] K is a field, often $K = \mathbb{R}$

[7] Skalski, Piotr. (2019). Gentle Dive into Math Behind Convolutional Neural Networks, URL: https://towardsdatascience.com/gentle-dive-into-math-behind-convolutional-neural-networks-79a07dd44cf9, June 22, 2020

[8] Common sizes are 2 x 2, 3 x 3, 5 x 5 and 7 x 7.

[9] Padding is almost always used to prevent information loss.

[10] The stride is usually set to one.

shrinkage to reduce computational effort. Large stride values decrease the amount of information that will be passed to the next layer. The most common way to optimize these hyperparameters is to use a validation set.

2.1.2 Activation functions und Biases

Before the results of a convolutional layer are loaded into the next layer, they usually go through a further intermediate step.

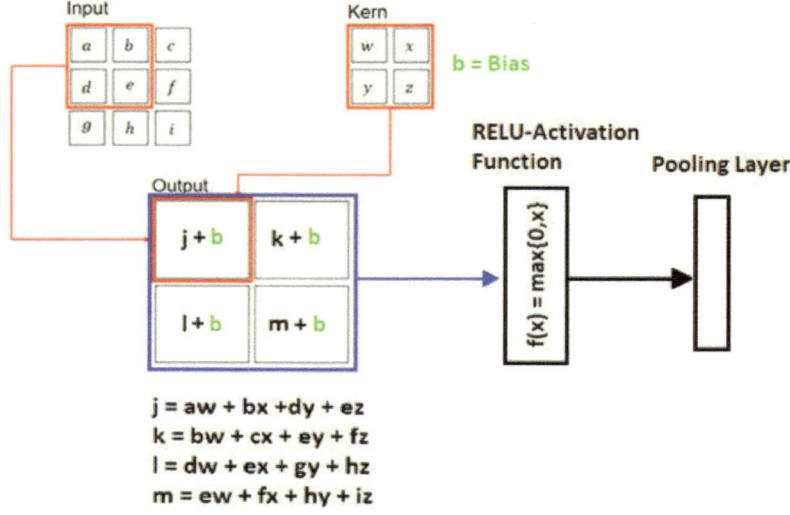

$$j = aw + bx + dy + ez$$
$$k = bw + cx + ey + fz$$
$$l = dw + ex + gy + hz$$
$$m = ew + fx + hy + iz$$

Figure 5: Bias and RELU-Activation Function[11]

Biases are assigned to each filter. In a ConvNet, the biases are added elementwise to the individual values of the result matrix. The sum is set into an activation function. It's function value serves as input for the pooling layer. ReLU or Sigmoid is commonly used as activation function.

---

[11] Based on: Nikolić, Zoran, (2019)

6

2.2 The Pooling Layer

A pooling layer aggregates the results of the convolutional layer. Its effect is to only pass only the most relevant signals from a given pixel width to the next layers. There are several pooling techniques to choose from. The most commonly used pooling techniques are *Max-Pooling* and *Average-Pooling*. For example, using a Max-Pooling layer, the highest value is selected and all others are discarded. Accordingly, the average of the convolutional values is used in average pooling. That pooling process not only reduces the amount of computational effort, but also protects against overfitting.

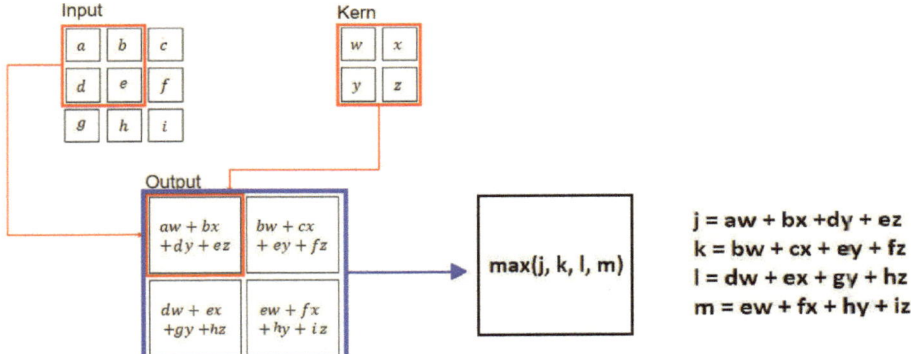

Figure 6: Pooling Operation[12]

For example, a 2 x 2 Max-Pooling layer reduces the 2 x 2 matrix result of a Convolutional layer to a single number (the maximum value). This provides a more abstract representation of the content and significantly reduces the amount of data to be processed.[13]

There are different approaches of ConvNets in which the exact sequence of Convolutional layers and Pooling layers strongly varies.[14]

After the last layer Pooling follow one or more Fully-Connected Layer.

[12] Based on: Nikolić, Zoran, (2019)

[13] Becker, Roland. (2019). Convolutional Neural Networks – Aufbau, Funktion und An-wendungsgebiete, URL: https://jaai.de/convolutional-neural-networks-cnn-aufbau-funktion-und-anwendungsgebiete-1691/, June 22, 2020

[14] Li, Z., Yang, W., Peng, S, Liu, F. (2020). A Survey of Convolutional Neural Networks: Analysis, Applications, and Prospects. Retrieved from University Library Archives, June 22, 2020

2.3　The Fully-Connected Layer

The fully connected layer has the structure of an ordinary neural network. That means, that all neurons are connected to all neurons of the previous and subsequent layer. In order to be able to load the outputs of the pooling layer into a fully connected layer, they firstly have to be rolled out. This process is known as *flattening*.

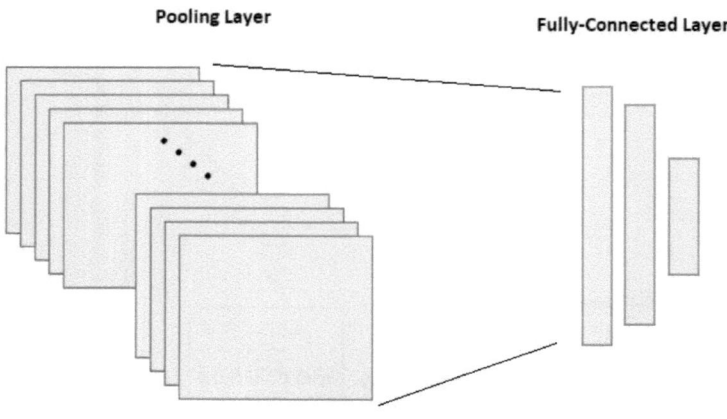

Figure 7: Flattening

The individual matrix values of the various pooling outputs are transformed into a common large input vector. This vector then serves as the input for the first layer of the neural network. According to ordinary neural networks, there may be additional hidden layers and an output layer. [15] In case of classification problems, this output layer usually has a softmax activation function. That is, the outputs of all neurons in the last layer add up to one and indicate the probability of class membership. The exact number of neurons in the output layer corresponds to the fixed number of classes. The edge weights between neurons are randomly chosen at the beginning and become further optimized during training phase. Stochastic gradient descent in combination with back propagation is the most commonly used training method. Furthermore, *categorical cross-entropy* is used to measure the error.

---

[15] Sometimes an additional dropout layer is used for regularization

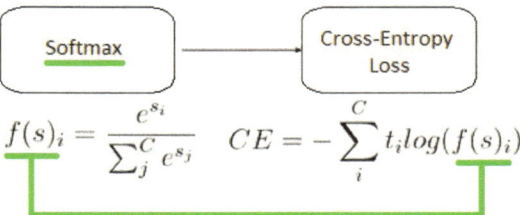

Figure 8: Categorical Cross-Entropy

It is the negative natural logarithm of the calculated probabilities for the respective class category. The loss function only records values of the softmax results, that ideally should be one. In order to calculate the error, a target vector (t) is used, that receives the optimal softmax result to correctly classify a given input, i.e. all entries are equal to zero except one.

2.4 Processing of colored images

As already mentioned at the beginning of the chapter, minimal changes need to be made if it is intended to process higher dimensional input like colored images. More precisely, the main difference is that three color channels are processed in parallel. First of all, the input is no longer a two-dimensional matrix. In gray-scaled images, each pixel is assigned only to a single integer value between 0 and 255. It indicates the brightness level, i.e. the brighter the pixel, the higher the value. A pixel in a colored image is assigned to three color values between 0 and 255 (one for each color). Therefore, colored images can be displayed as a three-dimensional vector matrix.

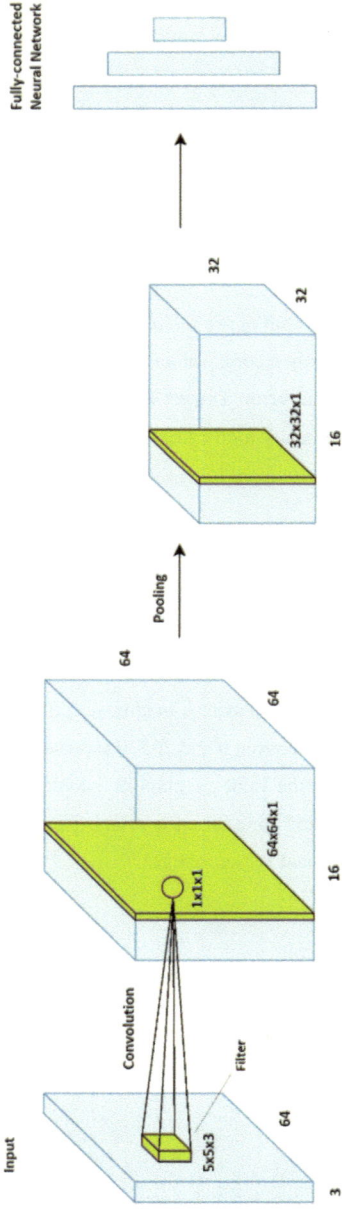

Figure 9: Processing of three-dimensional input data

Each color dimension can be treated in the same way as in case of gray-scaled images The only difference regarding calculation is that the individual point products are added up before playing them into the pooling layer. In this way, the three color dimensions are united in a single one. The individual calculations can be implemented efficiently. An individual three dimensional filter block (height, width and color channel) is used for each result matrix. It runs through the vector matrix of the input in the same way as in the gray-scaled case. The filter weights are shared again among same color channels. Compared to ordinary neural networks, that further enhances run time due to the effect of weight sharing. The pooling layers do not have to be adjusted, because the format of the pooling input does not change due to the adapted calculation in the Convolutional layer. The same applies to the fully-connected layer.

3 Advantages of Convolutional Neural Networks

3.1 Parameter Reduction

Compared to ordinary artificial neural networks, the main advantage of ConvNets is the reduction of parameters due to its architecture and functionality. That enables the network to be trained faster and save memory.

3.1.1 Weight Sharing in Convolutional Layers

Weight Sharing is used by ConvNets to detect particular features at any location in an image. It consists of having several connections controlled by a single weight. That is, a number of connections share the same weights saved in one moving filter.

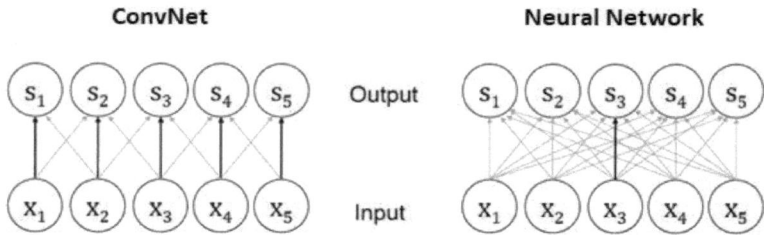

Figure 10: Weight Sharing[16]

Each entry of the filter is used at more than one position of the input. That weight sharing mechanism in ConvNets can drastically decrease the number of parameters.

3.1.2 Dimensionality Reduction via Pooling

Pooling layers are based on the principle of local image correlation to reduce the dimensionality of an image, that can reduce the amount of data while retaining useful information. It reduces the number of parameters by removing trivial features. This process is called *downsampling*. [17]

---

[16] Nikolić, Zoran (2019)
[17] Comp.: Li, Z. (2020)

3.2 Object Detection

Ordinary neural networks require vectors as input. Hence, the individual pixels of an image input must be flattened. Therefore, neural networks are unable to recognize objects in an image regardless of their position. For example, an image displaying a cat in the top left corner has a completely different input vector compared to the same image locating the cat in the bottom right corner. ConvNets are able to process input in matrix form without any transformations. There are already various examples of object detections methods using ConvNets.[18]

---

[18] Comp. Li, Z. et al. (2020)

4 Application to the MNIST Dataset

This is an application of ConvNets. The aim is to classify images from the MNIST data set. It consists of greyscale images that are handwritten digits (0-9). The data set is split into a training set of 60,000 images and a test set of 10,000 where each image is of 28 x 28 pixels in width and height.

Firstly, an ordinary neural network is used for classification. Subsequently, we try to enhance the performance by using additional Convolutional and Pooling layers. The performance of the two classification methods will be compared using the accuracy measure. We implement our methods in Python using the deep learning library Keras.

At first, the data is loaded. Then, the fully-connected neural network is built. The basic steps are the following:

- Flatten the input image dimensions (28 x 28) to an array (784 x 1)

- Normalize the image pixel values (divide by 255)

- Transform single valued true class labels to class vectors (10 x 1)

- Build the model architecture (Sequential) with Dense layers

- Train the model and make predictions based on test set

The neural network is composed of an input layer of length 784, a hidden layer with 100 neurons and ReLU activation function and an output layer with 10 neurons with soft-max activation function.

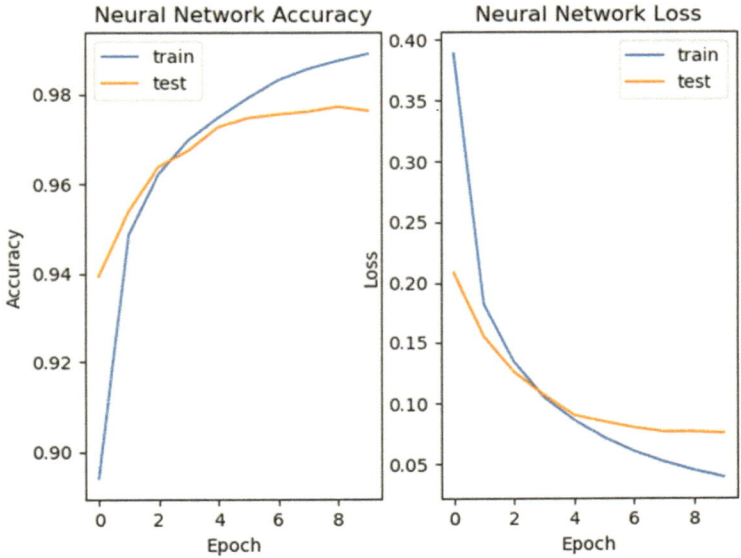

Figure 11: Accuracy and Loss of the Neural Network

The model was trained for 10 epochs using categorical cross-entropy loss and adam-optimizer. A batch size of 128 was used. As we can see in figure 11, the model achieved an accuracy of 0.9772 on the test set after nine epochs.

In the next step, we expand our existing model by additional layers. We add a single 2D-Convolutional layer followed by a Max-Pooling layer to the model. We use twenty-five 3 x 3 filters for the convolution. The stride is set to one. Furthermore, padding is not applied. ReLU serves as activation function for the additional layer. The rest of the settings remain the same as in the previous model.

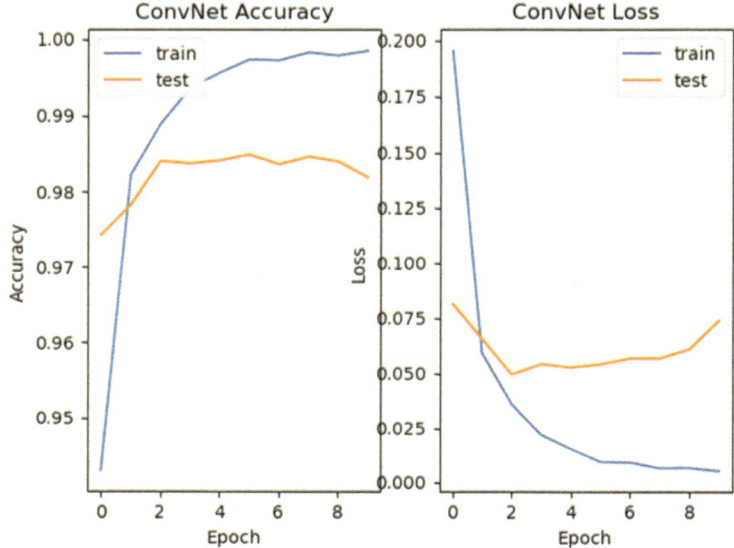

Figure 12: Accuracy and Loss of the ConvNet

Again, the model was trained for 10 epochs using categorical cross-entropy loss and adam-optimizer. A batch size of 128 was used. As figure 12 shows, the model accuracy increased to 0.9849 on the test set after six epochs.

5 Summary

ConvNets are one of the best learning algorithms for understanding image content. They can be used for image segmentation, classification, object detection and other non-image related tasks like text processing as well. At the beginning, Convolution supports the extraction of useful features from locally correlated data points. The output of the Convolutional layers is then assigned to the non-linear processing unit, called pooling layer. Finally, a small ordinary neural network is used to convert the output of those layers into final results. ConvNets are mainly used for high dimensional image data due to their decisive advantages. Compared to ordinary neural networks, the main advantage is the reduced number of parameters, that significantly accelerates the training process and saves storage space. Furthermore, they offer the possibility of object detection. The effective application of ConvNets can be demonstrated by numerous examples.

6 Literature

Le Cun, Y., Boser, B., Denker, J. S., Henderson, D., Howard, R. E., Hubbard,W., & Jackel, L. J. (1989). Backpropagation applied to handwritten zip code recognition. Neural Computation, 1, 541–551.

Li, Z., Yang, W., Peng, S, Liu, F. (2020). A Survey of Convolutional Neural Networks: Analysis, Applications, and Prospects. Retrieved from University Library Archives, June 22, 2020.

Li, Z., Yang, W., Peng, S, Liu, F. (2020). A Survey of Convolutional Neural Networks: Analysis, Applications, and Prospects. Retrieved from University Library Archives, June 22, 2020.

Becker, Roland. (2019). Convolutional Neural Networks – Aufbau, Funktion und Anwendungsgebiete, URL: https://jaai.de/convolutional-neural-networks-cnn-aufbau-funktion-und-anwendungsgebiete-1691/, June 22, 2020.

Nikolić, Zoran: (2019). Convolutional Neural Networks, URL: http://www.mi.uni-koeln.de/wp-znikolic/wp-content/uploads/2019/06/11-Odenthal.pdf, June 22, 2020.

Torres.AI, Jordi: (2018). Convolutional Neural Networks for Beginners, URL: https://towardsdatascience.com/convolutional-neural-networks-for-beginners-practical-guide-with-python-and-keras-dc688ea90dca, June 22, 2020.

Skalski, Piotr. (2019). Gentle Dive into Math Behind Convolutional Neural Networks, URL: https://towardsdatascience.com/gentle-dive-into-math-behind-convolutional-neural-networks-79a07dd44cf9, June 22, 2020.

Li, Fei-Fei, Krishna, Ranjey, Xu, Danfei, Byun, Amelie. (2020). Convolutional Neural Networks for Visual Recognition, URL: http://cs231n.stanford.edu/, June 22, 2020.

Tatan, Vincent. (2019). Understanding CNN (Convolutional Neural Network), URL: https://towardsdatascience.com/understanding-cnn-convolutional-neural-network-69fd626ee7d4, June 22, 2020.

7   Appendix

7.1   Python Code

The following code was used for chapter four:

```
from keras.datasets import mnist


# Data loading
(X_train, y_train), (X_test, y_test) = mnist.load_data()


# Shape printing
print("X_train shape", X_train.shape)
print("y_train shape", y_train.shape)
print("X_test shape", X_test.shape)
print("y_test shape", y_test.shape)


# Keras imports to build the fully-connected neural network
from keras.datasets import mnist
from keras.models import Sequential
from keras.layers import Dense
from keras.utils import np_utils


# Flattening
X_train = X_train.reshape(60000, 784)
X_test = X_test.reshape(10000, 784)
X_train = X_train.astype('float32')
X_test = X_test.astype('float32')


# Normalization
X_train /= 255
X_test /= 255
```

```
# Label Transformation
n_classes = 10
print("Shape before one-hot encoding: ", y_train.shape)
Y_train = np_utils.to_categorical(y_train, n_classes)
Y_test = np_utils.to_categorical(y_test, n_classes)
print("Shape after one-hot encoding: ", Y_train.shape)


# Building ordinary Neural Network
model = Sequential()
# Hidden layer
model.add(Dense(100, input_shape=(784,), activation='relu'))
# Output layer
model.add(Dense(10, activation='softmax'))


# Model summary
model.summary()
# Compiling
model.compile(loss='categorical_crossentropy', metrics=['accuracy'], optimizer='adam')
# Training and testing
import time
start = time.time()
history = model.fit(X_train, Y_train, batch_size=128, epochs=10, validation_data=(X_test, Y_test))
stop = time.time()
print(f"Training time: {stop - start}s")


# Plotting Histories
import matplotlib.pyplot as plt
```

```python
# Accuracy
plt.subplot(1, 2, 1)
plt.plot(history.history['accuracy'])
plt.plot(history.history['val_accuracy'])
plt.title('Neural Network Accuracy')
plt.ylabel('Accuracy')
plt.xlabel('Epoch')
plt.legend(['train', 'test'], loc='upper left')
# Loss
plt.subplot(1, 2, 2)
plt.plot(history.history['loss'])
plt.plot(history.history['val_loss'])
plt.title('Neural Network Loss')
plt.ylabel('Loss')
plt.xlabel('Epoch')
plt.legend(['train', 'test'], loc='upper right')
plt.show()


# Max Test Accuracy
print(max(history.history['val_accuracy']))
# Min Test Loss
print(min(history.history['val_loss']))


#-----------------------------------------------------------


# Further keras imports for extension to ConvNet
from keras.layers import Conv2D, MaxPool2D, Flatten

# loading the dataset
(X_train, y_train), (X_test, y_test) = mnist.load_data()
```

```python
# Reshape the input data
X_train = X_train.reshape(X_train.shape[0], 28, 28, 1)
X_test = X_test.reshape(X_test.shape[0], 28, 28, 1)
X_train = X_train.astype('float32')
X_test = X_test.astype('float32')


# Normalization
X_train /= 255
X_test /= 255


# Label transformation
n_classes = 10
print("Shape before one-hot encoding: ", y_train.shape)
Y_train = np_utils.to_categorical(y_train, n_classes)
Y_test = np_utils.to_categorical(y_test, n_classes)
print("Shape after one-hot encoding: ", Y_train.shape)


# Building ordinary Neural Network with additional Layers of ConNets
model = Sequential()
# Convolutional Layer
model.add(Conv2D(25, kernel_size=(3,3), strides=(1,1), padding='valid', activation='relu', input_shape=(28,28,1)))
# Pooling Layer
model.add(MaxPool2D(pool_size=(1,1)))
# flatten output of conv
model.add(Flatten())
# Hidden layer
model.add(Dense(100, activation='relu'))
# Output layer
model.add(Dense(10, activation='softmax'))
```

```
# Model summary
model.summary()
# Compiling
model.compile(loss='categorical_crossentropy', metrics=['accuracy'], optimizer='adam')
# Training and testing
start = time.time()
history = model.fit(X_train, Y_train, batch_size=128, epochs=10, validation_data=(X_test, Y_test))
stop = time.time()
print(f"Training time: {stop - start}s")


# Plotting Histories
# Accuracy
plt.subplot(1, 2, 1)
plt.plot(history.history['accuracy'])
plt.plot(history.history['val_accuracy'])
plt.title('ConvNet Accuracy')
plt.ylabel('Accuracy')
plt.xlabel('Epoch')
plt.legend(['train', 'test'], loc='upper left')
# Loss
plt.subplot(1, 2, 2)
plt.plot(history.history['loss'])
plt.plot(history.history['val_loss'])
plt.title('ConvNet Loss')
plt.ylabel('Loss')
plt.xlabel('Epoch')
plt.legend(['train', 'test'], loc='upper right')
plt.show()
```

```
# Max Test Accuracy
print(max(history.history['val_accuracy']))
# Min Test Loss
print(min(history.history['val_loss']))
```

YOUR KNOWLEDGE HAS VALUE